I0697329

Biology Revealed Uncovering the Architecture of Life

From Genes to Organisms: An Exploration of Biological Systems

Kitchen Mage

damage or risk resulting from the use and use, directly or indirectly, of any material given in this book is disclaimed by the publisher and the author(s).

The intended viewers

1. Majors in biology at the undergraduate level
2. Those who are interested in biology as hobbies
3. Instructors and learners in the biological sciences
4. Readers in general who are curious about the science of life
5. College students specializing in environmental science, genetics, or medicine.
6. Parents who homeschool and are searching for a thorough biology resource.
7. Biologists and other experts in the field looking for a summary or reminder.
8. those thinking about changing careers to ones in biology.
9. Science communicators search for easily comprehensible materials to convey biological concepts to a range of audiences.

Overview

Chapter 1: Life's Foundational Elements

Chapter 2: Genetics: Deciphering Life's Code

Chapter 3: The Story of Life Through Evolution

The Evolutionary Theory: The Legacy of Darwin

Natural Selection, Mutation, and Genetic Drift as Mechanisms of Evolution

Fossils, Comparative Anatomy, and Molecular Biology Provide Evidence for Evolution

Chapter 4: Ecology and Biodiversity

Examining Ecosystems: How Organisms Interact with Their Environment

Conservation of Biodiversity: The Range of Life on Earth Biology: Maintaining the Richness of Earth

In summary

Summary: Salutations from the Field of Biology

Greetings from the fascinating field of biology, where the details of the blueprint for life are revealed and investigated. Biology provides a doorway to comprehending the basic processes that form life in all its richness and complexity, from the huge ecosystems that cover our planet to the microscopic worlds of cells.

To understand the mechanics governing life itself, we will delve into the very fabric of existence on this journey, which is a voyage of discovery. We may understand the inner workings of all living things, from the tiniest microbes to the magnificent animals that inhabit the Earth, by looking through the lens of biology. Fundamentally, biology is not just a field of study but also a body of knowledge that supports Innumerable aspects of our lives. In addition to gaining a greater knowledge of the natural world, solving the mysteries of genetics, evolution, and ecological dynamics also provides us with priceless insights that guide scientific discoveries, conservation initiatives, and technological advancements.

Yes, there are benefits to knowing your life's blueprint that go well beyond the purview of academia. It gives us the ability to see how all living things are interconnected, to see how we fit into the complex web of life, and to accept the duty of stewardship towards our planet and its inhabitants.

As we set out on this thrilling voyage through the mysteries of biology, let us approach each finding with inquisitiveness, modesty, and a deep feeling of awe. Because every discovery we make reveals something new about the vastness and complexity of the world we live in. Greetings from the field of biology, where exploration and learning are ahead.

Chapter 1: The Basic Components of Life

We start our investigation of biology with the smallest unit of life, the cell. These minuscule organisms are the building blocks of all living things, possessing an astounding variety of structures and functions that are responsible for the wonders of life.

Analyzing Cells The fundamental units of living cells, or the building blocks of life, are intricate and precisely sized marvels that exist in a variety of shapes and sizes. Examining cells exposes the thread that unites all living things, from single-celled organisms to multicellular entities. Biologists have uncovered the complex machinery governing cellular life via painstaking observation and experimentation, providing light on the processes of growth, reproduction, and adaptation.

The composition and operation of cells

The study of structure and function is at the core of cellular biology. Many different organelles, each with a specific function to perform, work together in a symphony of activity within the boundaries of the cell membrane. From the ribosomes that synthesize proteins to the mitochondria that provide energy, the complex architecture of the cell is a reflection of a beautiful design maximized for robustness and efficiency.

The Role of Cells in Energy Production and DNA Repair

Many activities within the dynamic limits of the cell guarantee its survival and vitality. The creation of energy, which is accomplished through a complex dance of metabolic pathways, powers the machinery of life and keeps cells functioning. In the meantime, processes for DNA replication and repair maintain the genetic blueprint's integrity and guarantee the continuation of life across generations.

Discovering a world of astounding diversity and inventiveness, cellular biology is a complicated field that requires further investigation. The secrets to the continuation and evolution of life are contained within each cell, which is a monument to the creativity of nature. Building on the fundamental concepts revealed in our analysis of cells, we will continue our investigation into the fascinating field of biology in the upcoming chapters.

Chapter 2: Genetics: Unlocking the Secrets of Life

Within the complex field of biology, genetics stands out as a fundamental field that provides a deep understanding of the processes underlying the diversity of living things. The study of genetics has fundamentally changed our knowledge of inheritance, evolution, and the very nature of life itself. This understanding stems from the breakthrough work of Gregor Mendel and the innovative findings of contemporary molecular biologists.

Mendelian Genetics: Laws of Heredity
The work of Gregor Mendel, whose painstaking experiments with pea plants revealed the fundamental laws of heredity, stands at the start of modern genetics. Mendel developed the rules of segregation and independent assortment based on his studies of trait inheritance, providing the foundation for our comprehension of the inheritance of genes. Mendelian genetics

offers a timeless foundation for investigating the inheritance patterns of features across generations, from the simplicity of pea plants to the complexity of human genetics.

DNA's Structure and Function

The extraordinary molecule known as DNA (deoxyribonucleic acid) is central to the study of genetics. The blueprint for life is contained in its graceful double helix structure, which codes for the growth, development, and operation of all living things. Scientists have unraveled the mysteries of heredity, showing how genetic information is inherited from parents and passed down through the generations, by deciphering the structure and function of DNA.Characteristics to Genes: An Extensive Review of Heredity Heredity spans a wide range of phenomena, from the observable characteristics of an organism to the underlying genes that control their expression. The complexities of genotype and phenotype, gene regulation and expression, and the interaction between genetic and environmental factors in determining an organism's features are all examined through the lens of genetics.

The study of heredity continues to solve the puzzles surrounding the diversity and adaptation of life, starting with Mendel's peas and continuing with genome-wide association studies in contemporary genomics.

This chapter takes us on a journey into the core of genetics, exploring the mechanisms underlying trait inheritance and genetic information transmission. As we decipher the code of life's complexities, we acquire a greater understanding of the wonders of evolution and the astounding variety of life forms that inhabit our world.

Chapter 3: Evolution Tells the Story of Life

The story of evolution is the most compelling in the vast narrative of biology. This chapter examines how life on Earth has changed over time, from its modest origins to the astounding variety of living things that now call our planet home. We can solve the puzzles of speciation, adaptation, and the complex web of life that binds all living things together by using evolutionary theory as a lens.

The Evolutionary Theory: Darwin's Legacy

The revolutionary findings of Charles Darwin, whose groundbreaking essay "On the Origin of Species" laid the groundwork, are at the core of modern biology.

for the natural selection theory of evolution. The diversity of life forms and the patterns of change seen in living things over time are compellingly explained by Darwin's observations of variation, competition, and survival in the natural world. Generations of scientists have been motivated to investigate the mechanics and implications of evolutionary processes by his legacy, which remains a cornerstone of biological theory.

The Mechanisms of Evolution: Natural Selection, Mutation, and Genetic Drift Genetic drift, natural selection, and mutation interact dynamically to produce evolutionary change. Often called the "engine" of evolution, natural selection promotes the survival and procreation of individuals possessing beneficial features, resulting in the slow accumulation of adaptations within populations. In the meanwhile, mutations create genetic diversity, which serves as the foundation for the processes of natural selection. Evolutionary paths are further shaped by genetic drift, which is the random variations in allele frequencies within populations, especially in small or isolated populations.

Evolution is Supported by Comparative Anatomy, Molecular Biology, and Fossils

The molecules of life itself, the rocks, and the bones all bear witness to the tale of evolution. With their ability to record species origin and extinction as well as the transitions that have altered the history of life on Earth, fossils provide windows into the distant past. Comparative anatomy sheds light on the anatomical parallels and divergences between different organisms, offering insights into their evolutionary connections and common ancestry. The study of molecular biology provides information about DNA sequences and A molecular "fossil record" provided by genetic relationships validates the patterns found in the fossil and anatomical data.

We get a great grasp of the interdependence of life and the processes that have shaped the amazing diversity of living things as we delve into the complexities of evolutionary theory. The tale of evolution continues to astonish and amaze us, beckoning us to delve further into the evolutionary journey of life, from the modest beginnings of single-celled bacteria to the astounding array of species that populate our world today.

Ecology and Biodiversity in Chapter Four

This chapter takes us on a tour of the complex web of life that supports ecosystems and increases the diversity of life on Earth. We examine the basic ideas of ecology and the significance of conservation in protecting Earth's natural heritage, from the vibrant populations of species that live in every nook and cranny to the intricate web of ecological interactions that form our planet. Analyzing Ecosystems: How Living Things Engage with Their Environment Ecosystems, which comprise dynamic networks of species and their physical environments, are essential to the survival of our planet. This section explores the various habitats found around the world, ranging from lush Discover the various ways in which creatures interact with their surroundings, from barren deserts to rainforests. By using the ecological lens, we may better understand the complex interactions that exist between producers, consumers, and decomposers as well as the movement of nutrients and energy that keeps ecosystems alive.

Biodiversity Conservation: The Variety of Life on Earth

The foundation of life on Earth is biodiversity, which includes the incredibly wide range of organisms that call our planet home. However, pollution, habitat loss, and climate change pose hitherto unheard-of risks to biodiversity. This section addresses the difficulties in preserving the diverse array of life on Earth, examining the tactics and programs designed to safeguard endangered species. protecting important habitats and encouraging sustainable behaviors that protect biodiversity for the next generations.

Retaining Earth's Richness in Conclusion

We have a great duty to safeguard and maintain the biodiversity that keeps life on

Earth alive as stewards of the planet. We consider the interdependence of all living things and the significance of fostering a harmonious relationship with the natural world in this last section. We can make sure that the diversity of Earth's biological legacy is preserved for future generations by adhering to ecological principles and making a commitment to conservation activities.

This chapter delves deeply into ecology and biodiversity, providing a greater understanding of the complex relationships that unite all living organisms and the essential to protecting the natural resources of Earth. Let us take the lessons from this chapter with us as we continue our investigation of biology, working to safeguard and conserve the amazing variety of life that exists all around us.

Physiology: The Mechanisms by Which Organisms Function.

Chapter five: explores the complex systems that control how organisms behave, from the molecular level of cells to the integrated organ systems that support life. We examine the amazing adaptations and regulatory mechanisms that allow living things to flourish in a variety of settings and preserve internal stability in the face of changing circumstances via the prism of physiology.

Maintaining Stability in Living Systems: Homeostasis The delicate balance of an organism's internal state, known as homeostasis, is the foundation of physiological regulation. This section looks at how organisms stay stable in the face of changing environmental conditions like pH, temperature, and nutrient availability. We decipher the complex systems that allow animals to maintain balance and flourish in dynamic settings, from the feedback loops of hormone control to the cellular mechanisms of osmoregulation and thermoregulation. Understanding Adaptations for Survival by Examining the Physiology of Plants and Animals

Numerous physiological adaptations have been developed by plants and animals to enable them to survive and flourish in their specific environments. The physiological mechanisms underlying these adaptations are examined in this section, ranging from the respiratory system to the photosynthetic processes of plants.

animal nervous systems and circulatory systems. From deep-sea critters to desert succulents, comparative analysis provides insights into the variety of tactics used by organisms to address the challenges of their habitats.

We learn more about the incredible flexibility and tenacity of life on Earth as we navigate the intricacies of physiological systems. The workings of organisms, from the tiniest cells to the most complex organ systems, display a tapestry of intricacy and cunning refined over millions of years of evolution. We will expand on the fundamental understandings gained from our study of physiology in the upcoming chapters as we continue to delve into the wonders of biology.

Chapter 6: Behavior and Ecology

This chapter delves into the intriguing relationship between behavior and ecology, examining how creatures engage with both their surroundings and one another. We explore the deep relationships that define life on Earth and the significant effects of human activity on ecosystems around the world, from the fundamental instincts that drive individual behavior to the intricate dynamics of social communities.

Animal Behavior: From Adapted to Learned Patterns

Animal behavior is a broad category that includes learned behaviors that are gained through experience as well as instinctive behaviors that are encoded in genetic predispositions. This section examines the range of behavioral modifications displayed by animals, ranging from communication and reproductive techniques to migration and feeding. By disentangling the complex relationships between genes, environment, and behavior, ethology studies provide light on the evolutionary roots and ecological significance of animal behavior.

Social Interactions in Animal Communities

Animal communities are centered on social interactions, which mold population dynamics and impact individual behavior. This section looks at the various ways animals interact with each other, ranging from territoriality and dominance hierarchies to cooperative hunting and parental care. We can better appreciate the ecological functions that social behaviors play in preserving ecosystem balance as well as the adaptive significance of social behaviors when we see them through the lens of social ecology. Ecological Footprints: A Knowledge of How Human Activity Affects Ecosystems

The health and stability of ecosystems are greatly impacted by human activity, which also causes pollution, habitat degradation, and biodiversity loss on a global scale. This section tackles the ecological legacy of human civilization by looking at how resource exploitation, urbanization, and consuming habits have changed the environment. By studying sustainability and conservation biology, we investigate ways to lessen human effects on ecosystems and promote a healthy coexistence of human society and the natural world.

We obtain a greater understanding of the intricate relationship between ecosystem and behavior as we traverse it. the intricate balance that keeps life on Earth going, as well as the interconnection of all living things. We'll be delving further into the complex web of life in the upcoming chapters, drawing on the knowledge gained from studying behavior and ecology to enhance our comprehension of the wonders of biology.

Chapter 7: Applications of Biology

This chapter examines the numerous applications of biological principles and findings to real-world problems, enhancing human health, and influencing global trends. We explore the profound effects of biological research and innovation on society, from using biotechnology to improve agriculture and industry to furthering medical advancement through novel treatments and therapies.

Biotechnology: Transforming Our Way of Life Benefit

Biotechnology is the broad umbrella term for a variety of methods and applications that use biological processes to improve human health and solve problems. This section looks at the many uses of biotechnology, ranging from environmental remediation and agricultural biotechnology to synthetic biology and genetic engineering. Biotechnologists open up new possibilities for increasing crop yields, creating innovative materials, and solving environmental problems by tinkering with DNA, proteins, and biological systems. This creates the foundation for a more robust and sustainable future.

Medical Development: Vaccinations to Gene Editing

The medical area is at the forefront of biological innovation, where discoveries in knowledge and treating diseases have transformed human health and longevity. In this section, we explore the remarkable progress achieved in medical science, from the development of vaccines and antibiotics to the advent of precision medicine and gene editing technologies. Through the application of molecular biology, genetics, and biotechnology, researchers and clinicians work tirelessly to combat diseases, alleviate suffering, and enhance the quality of life for individuals around the globe.

Ethics in Biological Research and Innovation: A Framework for Thought

As we push the boundaries of biological knowledge and technological capabilities, we are confronted with profound ethical questions and considerations. In this section, we delve into the ethical implications of biological research and innovation, exploring the principles and frameworks that guide responsible conduct and making choices. With humility, integrity, and a dedication to protecting the rights and dignity of every person, we traverse the challenging field of bioethics, addressing everything from informed consent and genetic privacy to equity and social justice issues.

As we come to the end of our investigation into the applications of biology, we are aware of the revolutionary potential of science to enhance lives, spur advancement, and influence global trends. By exercising good stewardship and carefully considering ethical issues, we may fully utilize biological research and innovation to build a more sustainable, just, and healthy world for future generations.

In conclusion: Reflecting on Our Biology Journey

We pause as we approach the end of our investigation into the fascinating field of biology to consider the vast array of information and understanding that has been revealed to us. We have traveled through the glories and complexities of life itself, from the microscopic worlds of cells to the vast ecosystems that shape our globe, and we have gained insights into the fundamental principles that control all living things.

We have come across the graceful laws of genetics, the transforming force of evolution, and the complex interactions between living things and their surroundings along the way. We have been astounded by the flexibility and tenacity of life forms, and we have seen the significant influence of human activities on ecosystems and grappled with ethical questions and considerations at the forefront of biological research and innovation.

Yet amidst the vast expanse of biological knowledge, one truth remains abundantly clear: the wonder and complexity of life are inexhaustible. As we continue our quest for understanding, may we carry forward the spirit of curiosity, humility, and reverence for the natural world that has guided us thus far. For in the pursuit of knowledge lies the key to unlocking the mysteries of existence and embracing the interconnectedness of all living beings.

Appendix: Glossary of Important Terms

To aid in further exploration and comprehension, we have compiled a comprehensive glossary of key terms encountered throughout our journey. This glossary ensures readers' comprehension at every learning step by offering definitions and explanations for terms ranging from cellular biology to ecological principles. It is an invaluable reference tool.

Further Resources for Further Investigation

We have selected several supplementary materials to help individuals who are keen to learn more about particular subjects or begin their biological discoveries. These resources, which range from suggested reading lists to online databases and instructional websites, provide a multitude of knowledge and perspectives to encourage ongoing biological research and learning.

As we say goodbye to this leg of our journey, let's continue the understanding, inquisitiveness, and gratitude for life's marvels that we have developed throughout time. May we continue to investigate the mysteries of biology with fervor, diligence, and a firm commitment to comprehending and protecting the diversity of life on Earth, whether as learners, teachers, researchers, or just inquisitive minds.